Oil Machines

Copyright © 1979, Raintree Publishers Limited

Library of Congress Number: 78-26333

  2   3   4   5   6   7   8   9   0   83   82   81   80

Printed in the United States of America.

Library of Congress Cataloging in Publication Data

Pick, Christopher C.
  Oil machines.

  SUMMARY: Describes various pieces of oil
machinery and discusses their uses.
    1.   Oil hydraulic machinery — Juvenile literature.
[1.   Oil hydraulic machinery.   2.   Machinery]
I.   Title.
TJ843.P52   621.2'04'24   78-26333
ISBN 0-8172-1327-9 lib. bdg.

Cover illustration: Jerry Scott

Photographs appear through the courtesy of the following
  companies:

Amoco (U.K.) Exploration Company pp. 8, 13 (bottom),
  14 (bottom), 16, 21,
Amoco (U.K.) Ltd: pp. 28, 29 (bottom)
Conoco: pp. 9, 10, 11, 17
Gulf Oil Corporation: pp. 3, 12, 13 (top), 14 (top),
  15, 22, 25 (top), 29 (top)
Shell International Petroleum Ltd: pp. 4, 5, 6, 7,
  18, 19, 20, 23, 25 (bottom), 26, 27
Sohio: p. 24

oil MACHINES

Christopher C. Pick

RAINTREE CHILDRENS BOOKS
Milwaukee • Toronto • Melbourne • London

Oil is very hard to find. It is in pools deep in the earth. The pools are called oil wells. They may be under the land or the sea. A group of wells close together is called an oil field.

Oil companies send people who study rocks to search for oil. They look at the ground very carefully. Rocks help to tell them whether oil might be underground.

If these experts think oil is there, they set off an underground explosion. It is like a small earthquake. Special instruments listen to what happens to the rocks underground. This is called a seismic survey.

This machine does a seismic survey. It sends an explosion through the water and into the seabed.

After the explosion, the experts study their instruments. They try to decide if there is oil.

If the chances look good for finding oil, a deep hole is dug. This is the only way to be sure oil is there.

The oil workers bring a machine called a drilling rig. The rig has a tall tower. The tower is called a derrick.

The rig workers use a drilling bit to cut through the rock. They join the bit to a pipe. This pipe is called a drill string. It hangs in the derrick.

Engines drive the drilling bit
into the earth. As it goes down,
more pipes are added to the
drill string.

If the rocks in the earth are
very hard, they wear out the bit. It
must then be changed often.

As the drilling bit goes deeper and deeper, mud and rock come up the drill string. If the drillers find oil, it rushes up the drill string too.

But sometimes pumps are needed to help get the oil up from the well.

After the well is finished, the oil is carried away from the well in pipelines. On top of each well is a special machine called a Christmas tree. Oil goes from the well through the Christmas tree and into the pipeline. The Christmas tree controls the amount of oil that goes into the pipeline.

To get oil from the sea, special rigs are needed. The jack-up rig is used in shallow water. Its legs sit on the bottom of the sea. It is called a jack-up rig because workers can make its legs taller or shorter.

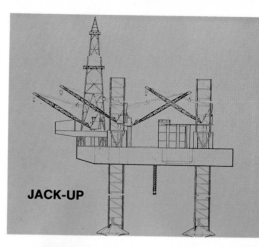

JACK-UP

This rig is used in deep water. It is called a semi-submersible rig. It has special legs filled with air. The rig can float to the drilling site. When it arrives, it is held in place with anchors. The rig then goes lower into the sea, ready for drilling.

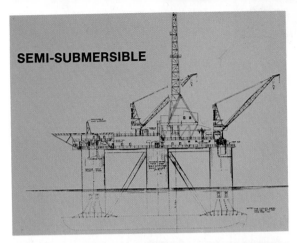

SEMI-SUBMERSIBLE

This rig is called a drillship. It
has a derrick built on its deck. The
drilling bit and the drill string go
down into the water through a hole
in the bottom of the ship.

Many people work on drilling rigs. Most of them help to drill for oil. Some do the cooking and make things comfortable for the others. Rig workers may even have time to catch fish from their rig.

It is not easy to get oil from a
well under the sea. First, a
production platform must be built.
The base of this platform was built
on land. It was pulled out to sea on
its side. Then it was turned over
and set on the seabed.

Next, the production platform is built up. Decks are made. In this picture, a crane is lifting a block of buildings on to the platform. The oil workers will live in these buildings.

This is another kind of
platform. Almost all of it was built
in a shipyard. Tugboats pull it
out to sea. When it has been set up
on the seabed, the legs will be
mainly underwater.

Helicopters often travel to
production platforms. They bring
people, food, and equipment to
the platforms.

The production platform is like a factory. It has everything needed to get oil up from the well. Oil workers may live on the platform for many days. The platforms must be strong enough to stand bad weather and rough seas.

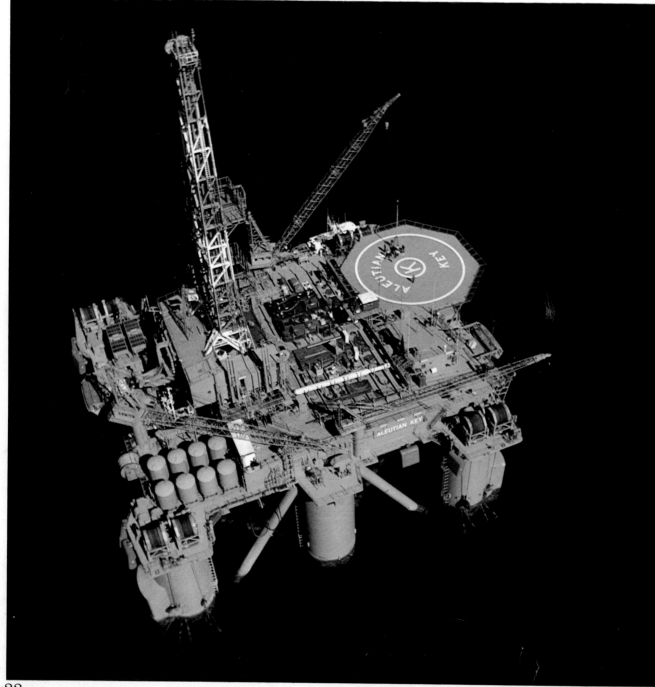

When the platform is ready, oil can be taken from the wells. A platform may be able to get oil from as many as 48 wells at the same time.

Oil that comes from a well is called crude oil. It cannot be used yet. First it must go to a refinery. The crude oil may travel through pipelines in the ground or in the sea.

Sometimes, big ships called oil tankers carry oil to a refinery.

This tanker is taking oil from a special storage tank in the sea. The oil was carried to the tank from an oil field.

The oil is unloaded at a
terminal. The oil stays in the
terminal until the refinery needs it.
Often the terminal is part of
the refinery.

The refinery changes crude oil into oil that we can use. It may become gasoline or kerosene. Some of these tanks hold crude oil. Others keep oil that has been refined.

When the refined oil is ready, it
is put into road tankers. These are
big trucks specially built to carry
oil. They take it wherever it
is needed.

Gasoline and diesel fuel are sold to motorists at service stations. At airports, kerosene is loaded on to airplanes.

Oil is also used in homes, on farms, and by factories.

GLOSSARY

Christmas tree	A special machine used on an oil well. The Christmas tree controls the amount of oil that goes from the well to the pipeline.
crude oil	The oil that comes from an oil well. Crude oil must be refined before it can be used.
derrick	The tall tower of a drilling rig.
drilling bit	A machine used to cut through rock and earth.
drilling rig	A machine that is used to get oil from under the ground.
drillship	A special oil rig that has a derrick on its deck. The drill string goes into the water through a hole in the bottom of the ship.
drill string	Pipes that are joined to a drilling bit. The drill string hangs in the derrick. Oil comes up through the drill string.
gasoline	A fuel that is made from oil.
jack-up rig	An oil rig used in shallow water to get oil from the ground under the sea.
kerosene	A fuel that is made from oil.
oil	A liquid found deep in the earth. Oil is used to make fuels and many other things.
oil field	A group of oil wells close together.
oil tanker	A ship that can carry very large amounts of oil.
oil well	A pool of oil that is deep underground.
pipeline	Long pipes that are used to carry oil away from the oil well.
production platform	A building used to get oil from wells under the sea. Workers can live on a production platform.
pump	A machine that is sometimes used to help get oil up from a well.
refinery	The place where crude oil is made ready so that it can be used.
seismic survey	A way scientists have to study the earth to help decide if there is oil under the ground.
semi-submersible rig	An oil rig used to get oil from the deep sea.
tugboat	A boat that can pull very heavy things over the water.

INDEX